ZAP IT!

Exciting Electricity Activities

U.S. edition published in 1999 by Lerner Publications Company,
by arrangement with Evans Brothers Limited, London, England.

Lerner Publications Company
A Division of Lerner Publishing Group
241 First Avenue North
Minneapolis, MN 55401

Website address: www.lernerbooks.com

Library of Congress Cataloging-in-Publication Data

Good, Keith.
Zap it! : exciting electricity activities / by Keith Good.
p. cm. — (Design it!)
Includes index.
Summary: Introduces the basics behind the operation of various battery-powered electrical devices and provides instructions for making some simple devices that use different kinds of circuits and switches.
ISBN 0-8225-3565-3 (lib. bdg. : alk. paper)
1. Electric apparatus and appliances—Juvenile literature.
2. Electricity—Experiments—Juvenile literature. [1. Electric apparatus and appliances.] I. Series.
TK148.G643 1999
621.31'042—DC21 99-36675

Printed in Hong Kong
Bound in the United States of America
1 2 3 4 5 6 - OS - 04 03 02 01 00 99

DESIGN IT!

ZAP IT!

Exciting Electricity Activities

Keith Good

Lerner Publications Company • Minneapolis

About this book

About this series

This series involves young people in designing and making their own working technology projects, using readily available salvaged or cheap materials. Each project is based on a "recipe" that promotes success and stimulates the reader's own ideas. The "recipes" also provide a good introduction to important technology in everyday life. The projects can be developed to different levels of sophistication according to readers' ability and can reflect their other interests. The series teaches skills and knowledge in a fun way and encourages creative, innovative ideas.

About this book

One reason for giving young people hands-on experience with electric circuits is that they are surrounded by electrical devices every day and practical experience helps them to understand their world better. This book focuses on making a range of battery-powered projects with an emphasis on fun, creativity, and readily available materials. The ideas can also be used to enhance other design projects, perhaps based on mechanisms, such as lights for a new fairground ride.

It is a good idea to observe the convention of RED wire for positive (+) current and BLACK wire for negative (-), as some components will only work if connected the right way. When working in groups, one or two batteries can be used by all to test their projects. Limiting the number of batteries saves cost - and reduces the noise from buzzers! Agreeing to keep the batteries in fixed places saves searching.

Safety

- Young people must be made to understand that household electricity can be lethal and that they must never tamper with it. This should be part of their general "keeping safe" education but always stressed when introducing electric circuits. All projects in this book use battery power only.

- Do not use rechargeable batteries. These give up a lot of energy (heat) quickly if short-circuited, for example, if a paper clip bridges the two terminals.

- Adult use of craft knives is strongly advised. These must be used with a cutting board or mat and a safety rule with a groove to protect the fingers. Cardboard is often best cut by an adult on a paper cutter with a guarded-wheel cutter. A sharp paper cutter is a good way to cut foil strips.

- Use wire strippers, not craft knives, to remove the plastic insulation from wires. If possible, try (and evaluate) different types of wire strippers before buying one.

Contents

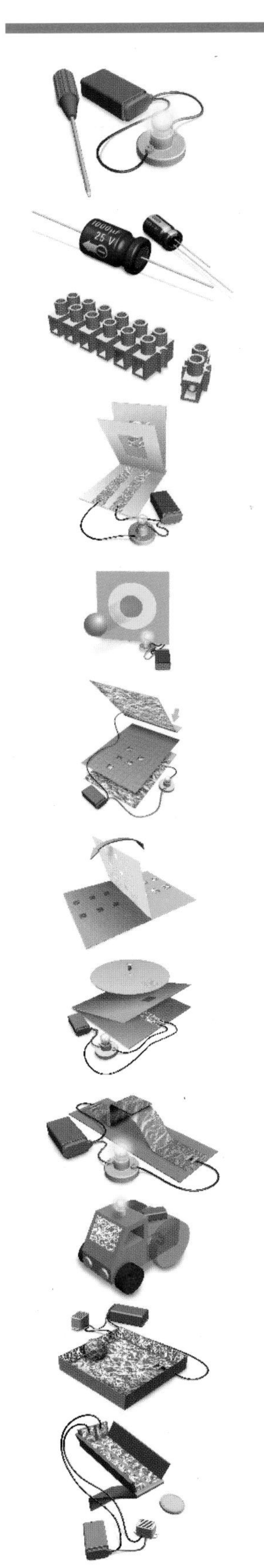

Electric circuits

Making series and parallel circuits

It is hard to imagine the world around us without electric power. We use it to cook our food and to keep it cool, light our homes, and power our computers and toys. How many other uses for electricity can you think of?

The activities on these two pages will help you with the projects in this book.

Safety

Remember that the projects in this book are all powered by batteries. The electricity that comes from the outlets in your house is much too powerful and could kill you. Use battery power only!

You will need

- a battery and snap connector
- 2 6-volt bulbs and 2 bulb holders
- multi-strand electrical wire (the kind with lots of thin wires inside the plastic)
- a wire-stripping tool
- a small screwdriver

What to do

1. Screw a bulb into a bulb holder.

2. Use your screwdriver to attach the battery connector to the bulb holder. Loosen the screws and loop one wire around each, then tighten the screws.

3. Snap on the battery, and the bulb will light.

Electricity is flowing from the battery through the wires and bulb and back to the battery. You have made a *series circuit.*

A series circuit

battery

connector

bulb and bulb holder

The series circuit you have made can be drawn like this:

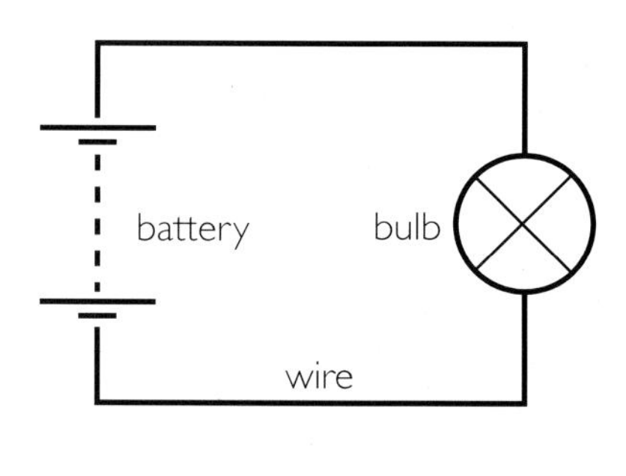

4. Undo one of the wires attached to the bulb. What happens?

A complete "loop" is needed for a circuit to work. A switch is a gap in a circuit that can be closed easily to turn things on. All the switches in this book work by closing a gap in a circuit.

5. Try adding another bulb to your series circuit to see what difference this makes.

Connect the bulbs with a short piece of wire. Use the wire stripper to remove the plastic insulation from each end of the wire first.

6. Now try making a *parallel circuit* (see below). Are the bulbs brighter or dimmer than before? A parallel circuit will be needed when your project includes more than one bulb, buzzer, or motor.

A series circuit with two bulbs

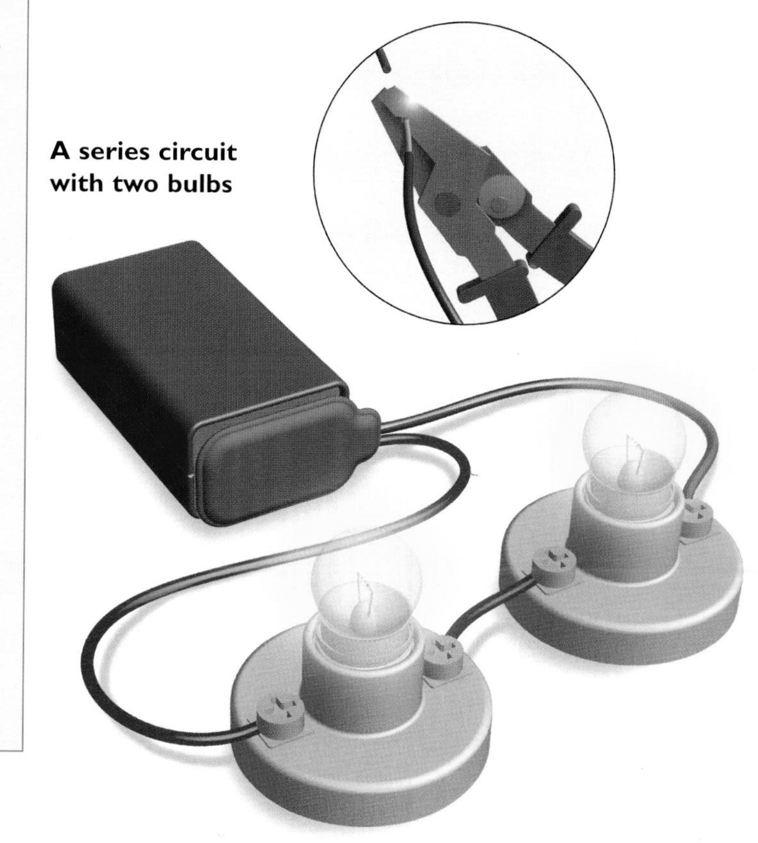

The series circuit with two bulbs can be drawn like this:

A parallel circuit

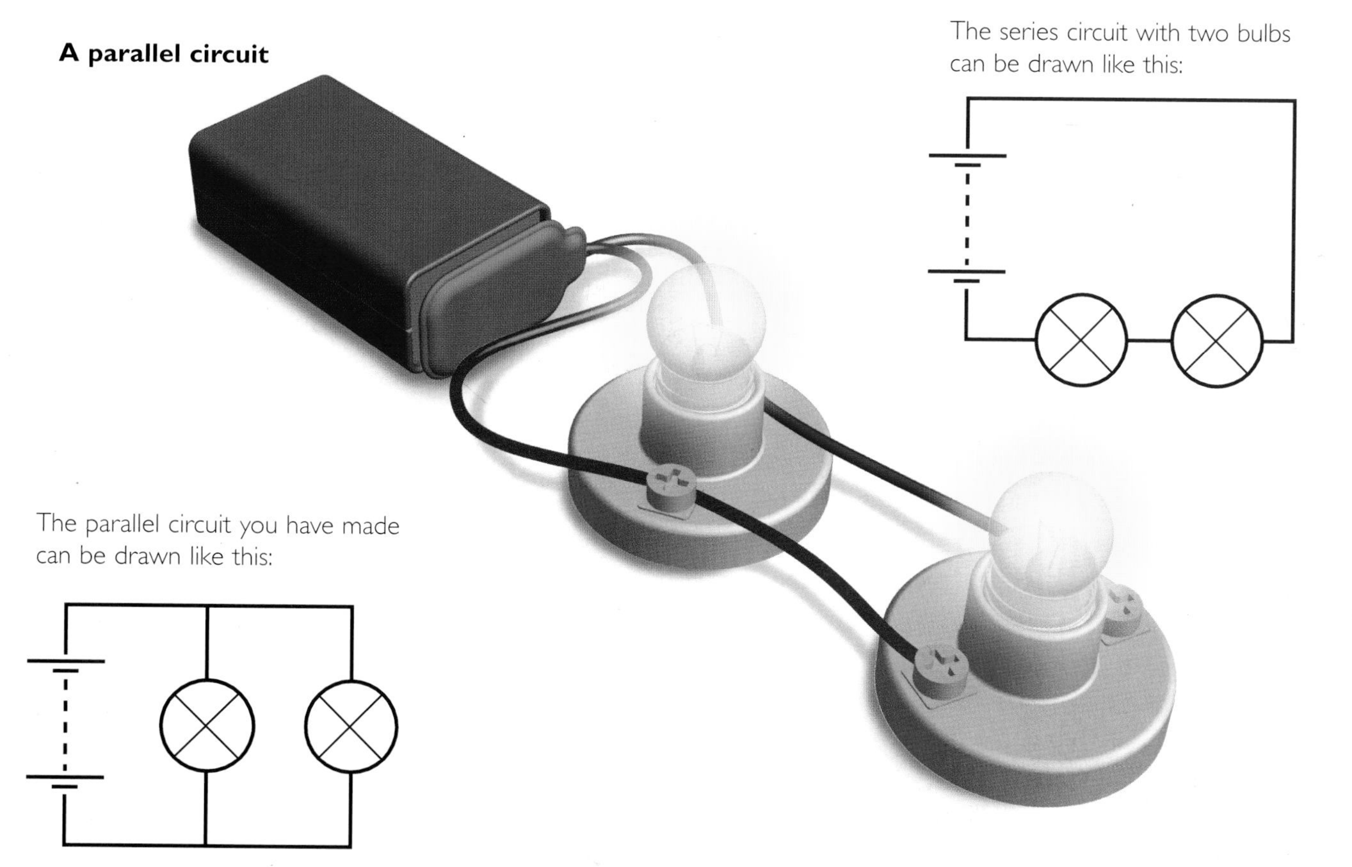

The parallel circuit you have made can be drawn like this:

Light, sound, and movement

Designing circuits for your projects

Electrical and electronic parts are often called *components.* Many different components are used in the personal stereos, computer games, and other products we see every day. The microchip (see photograph, right), used in computers and many other devices, is one component that has changed our lives.

Here are some simpler components that you can use in circuits and design projects.

Light

Bulbs are useful when you want very bright lights or to light something up. You can make them even brighter by making a reflector (which flashlights and headlights use) out of foil. Bulb holders are quite bulky, but they can often be hidden inside a project while leaving the bulb showing.

LEDs (light emitting diodes) are the small red, green, or yellow lights on computers, phones, and many other products. How many LEDs can you find in your home? LEDs are cheap, colorful, and give projects a "high tech" look.

The short leg of the LED must be attached to the black (negative/–) wire. The flat part of the LED is always next to the short leg. Attach the long leg to the red (positive/+) wire.

LEDs can cope with 3 volts, but if you are using a 9-volt (9V) battery you must add a resistor or the LED will be ruined. You need a 200 ohm resistor. It can be attached to either leg of the LED.

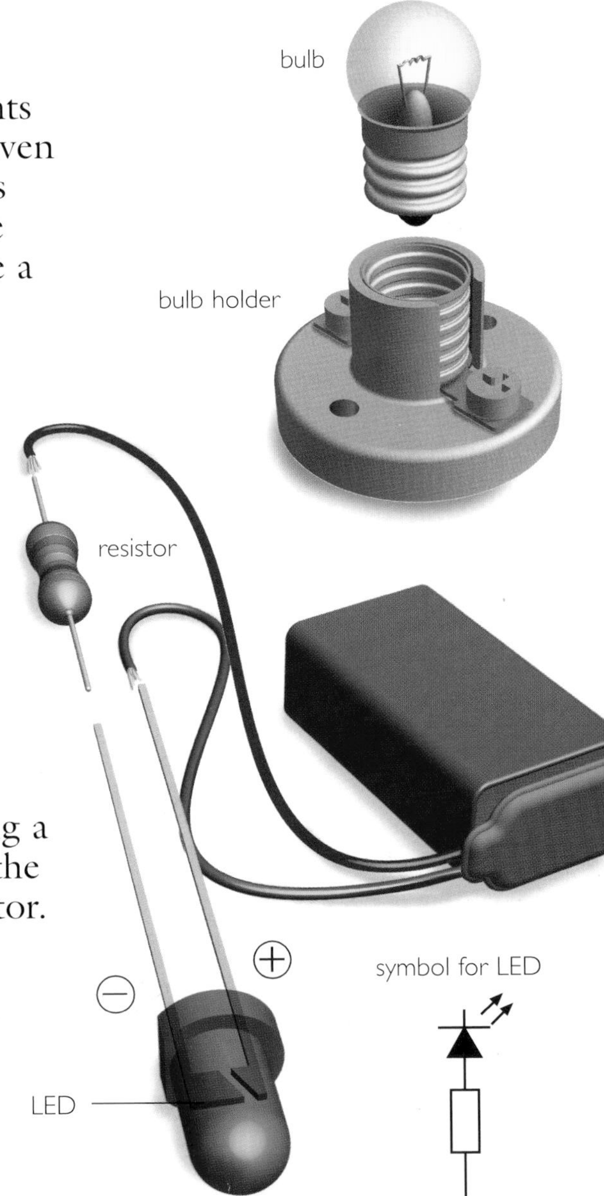

Sound

Round buzzers and rectangular buzzers sound different! The red wire from the buzzer must be joined to the red wire from the battery. The black wire from the buzzer must be joined to the black wire from the battery.

You can use a capacitor to make a buzzer or light stay on for a short time (then fade) after the switch goes off. The capacitor must be connected the right way. Look for the negative (–) markings that show you where to connect the black wire from the battery.

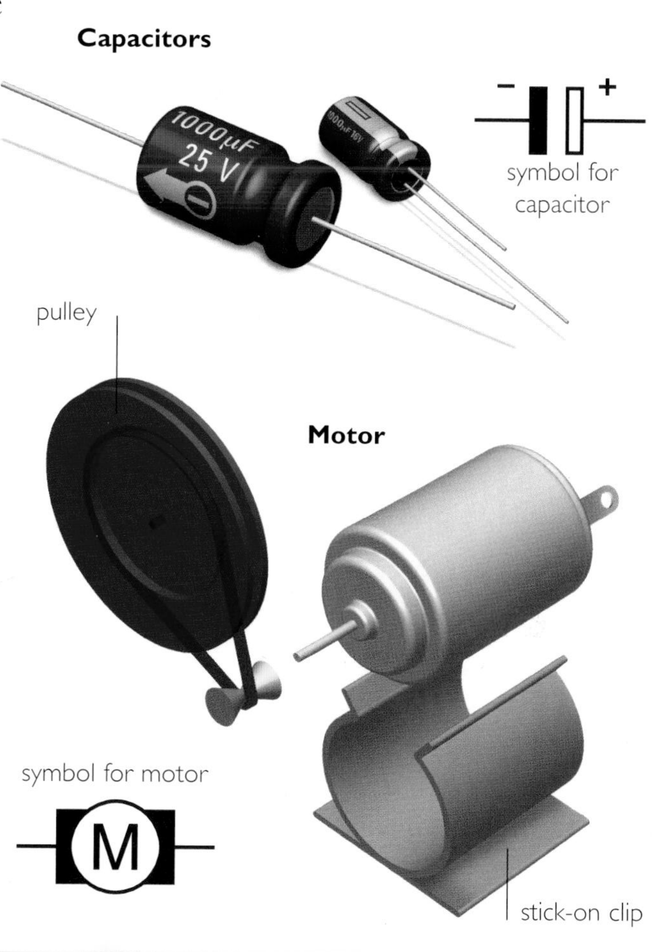

Movement

Motors often turn too fast and not strongly enough for many projects. Pulleys or gears are added to reduce the speed and increase the force of the motor.

To do useful work, motors need to be held in place with stick-on clips.

Tip

Bulb holders and buzzers can be attached to your work with plastic bag ties or single-strand wire.

Getting ideas

You could design and model your own electronic product of the future – it doesn't really have to work! Make a list of battery-powered products to help you get ideas. Look at battery-powered products and, **with adult help**, look inside broken, unwanted ones.

Shopping catalogs and visits to stores will also give you ideas for your own designs. You could include some of the components on these two pages and control them with switches from this book.

More about making circuits

Connecting wires and adding switches

Circuits only work when the parts are connected properly. Solder – a mixture of tin and lead – can be melted between the parts, using a hot soldering iron. This makes a good strong joint that conducts electricity well, but soldering can be dangerous. Here are some other ways to make your circuits.

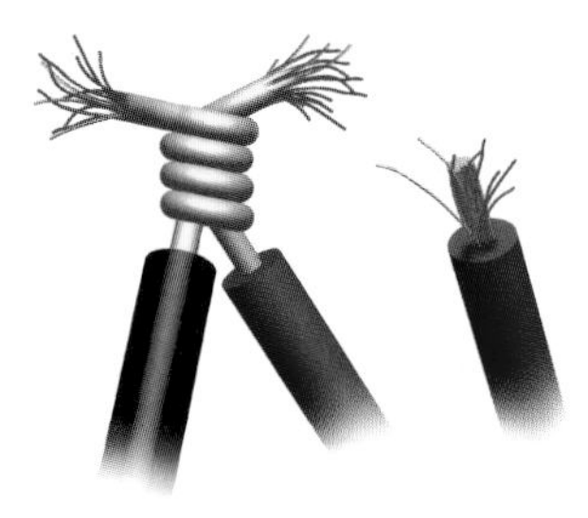

Multi-strand wire

1. Use wires with lots of strands. They are easiest to twist together tightly. Strip off about half an inch of the insulation first. Putting electrical tape over the connections can help to keep them together and prevent accidental electrical connections.

2. Although connecting strip is bulky, it can be a useful way to join parts. Cut off what you need and attach the wires with a screwdriver.

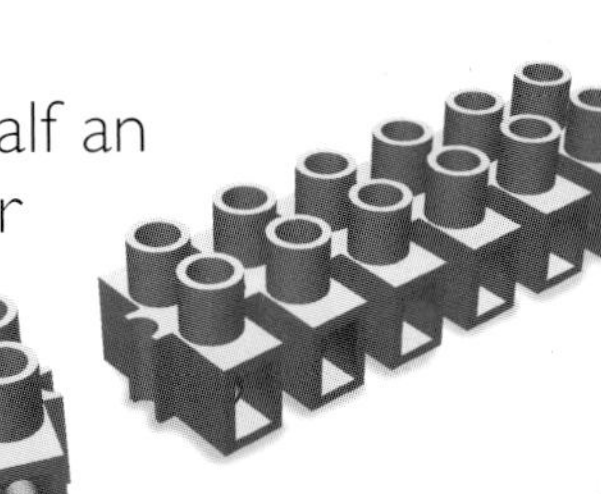

Connecting strip

3. Making a circuit on a board can help you to see what you are doing and avoid mistakes. Stick the wires down with masking tape and test your circuit before connecting it to a project. Remember to make the wires long enough.

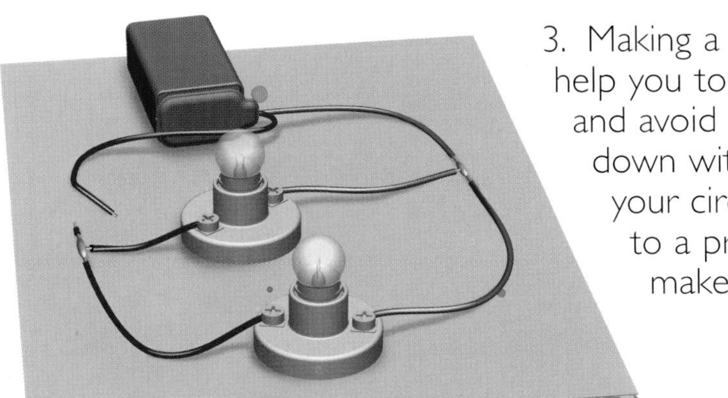

Where you put your switch determines which components it will control. If the switch is put in this position, it controls both components.

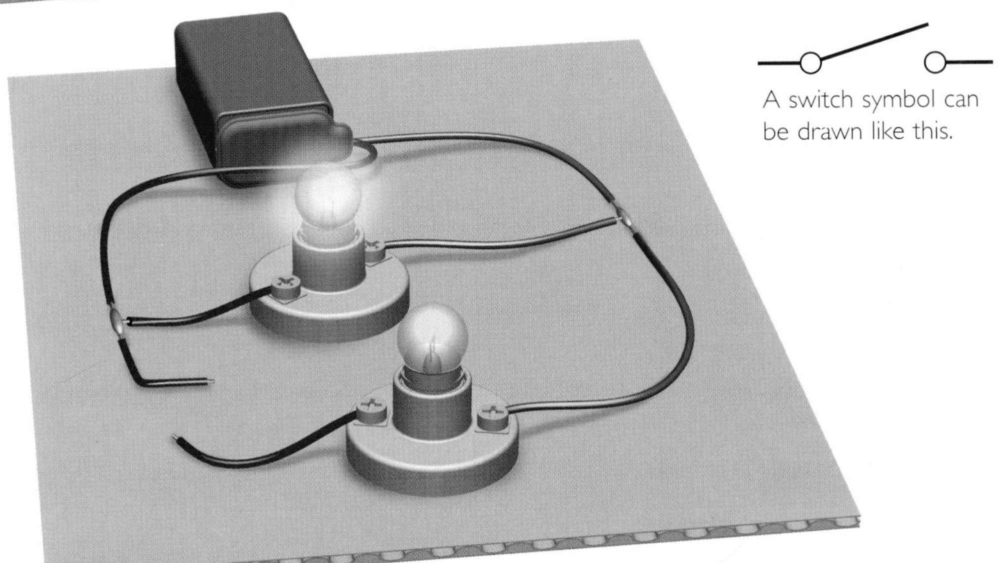

A switch symbol can be drawn like this.

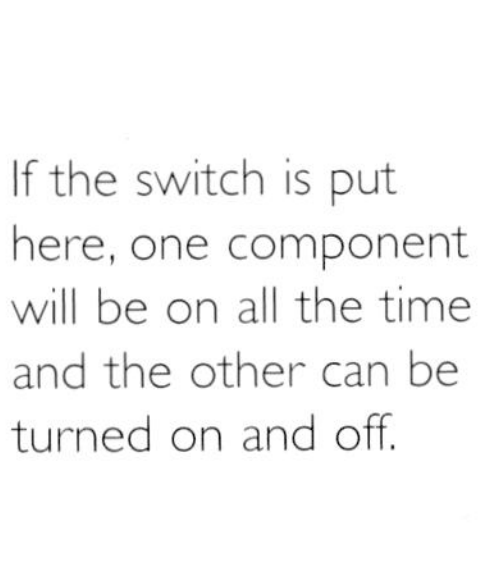

If the switch is put here, one component will be on all the time and the other can be turned on and off.

This is where to put a capacitor (see page 9) so that the buzzer stays on for a while and gradually fades after the switch is off.

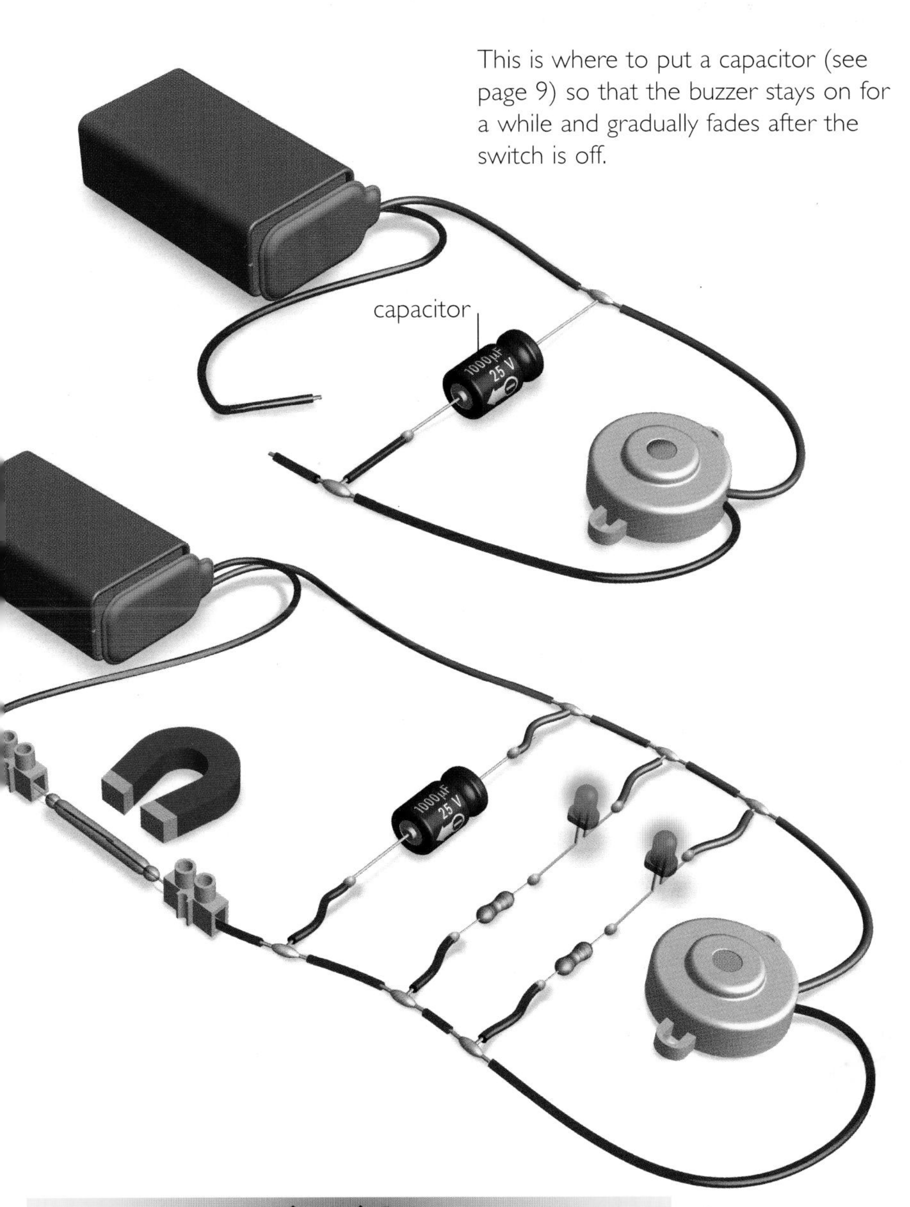

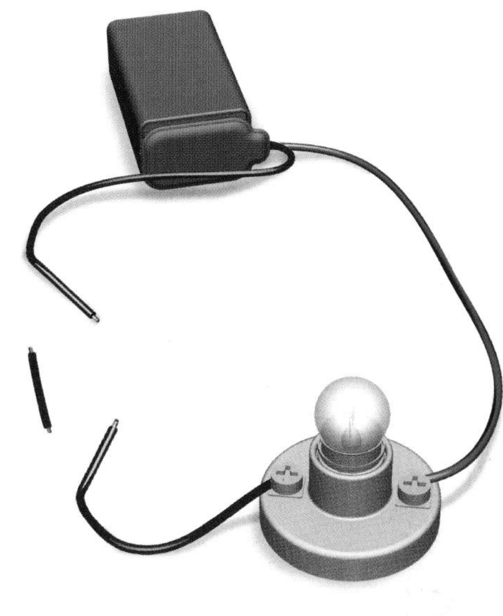

Here both switches must be on before the circuit will work. One switch (like a pressure pad switch, page 12, or a reed switch, page 24) could be hidden to work like a secret key.

When a magnet is passed over the reed switch (page 24), the LEDs (page 8) glow, then slowly dim. The buzzer goes on and makes a wailing noise before becoming quiet. This circuit could be used to make a creature that "comes to life" when it is stroked.

Getting ideas

Try to think of different uses for the circuits shown on these pages. Draw some other circuits that you think would work. You could draw lifelike components or use symbols. (See pages 6-9.) Try the circuits out if you can.

If you can borrow different kinds of wire strippers, try them out and compare them. Which do you prefer and why? **Safety**: Some wire strippers have sharp parts, so be careful.

Tip

What to do if your circuit doesn't work:

- Check that the battery is working. Try it on a spare buzzer or light.
- Check that all the parts are in the right place and connected the right way.
- If a circuit that used to work stops working, see if something has come loose.

Pressure pads

Making your own pressure pad switch

Pressure pad switches (sometimes called membrane panel switches) are made from thin layers. They are tough and don't take up much room. Look for pressure pads on cash machines, microwave ovens, and other machines that we use every day.

You can make a pressure pad switch to control anything that is battery-powered.

kitchen foil

staple

You will need

- 3 3" by 5" index cards
- scissors
- a glue stick
- a stapler
- kitchen foil
- something to switch on, such as a bulb, buzzer, or LED (see page 9)

What to do

1. Cut a hole in one index card like the one in the picture (page 12). This will be the middle layer.

2. Use the hole to stencil a square on the top and bottom layers.

3. Stick a square of foil over the square on the top layer.

4. Stick two strips of foil *across* the square on the bottom layer, as shown on page 12.

5. Staple your switch together at one end like a book.

6. Staple the bare wires tightly to the foil strips - and try it!

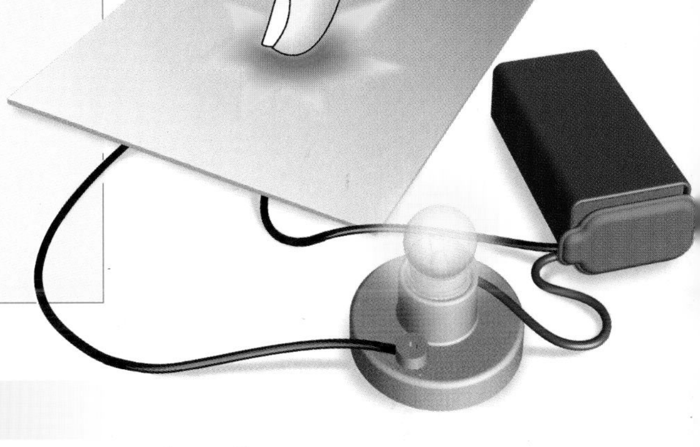

Getting ideas

How many ways are there to turn a pressure pad on? Think about the following questions to help you get ideas. Could you swing something at it, put it where it would be turned on by accident, sit on it, or roll something over it? Could you drop something on it, stand on it, press it with your finger, or do something else?

What could you turn on with a battery-powered pressure pad – a light bulb, an electric motor, an LED, a buzzer, or something else?

Could you design a creature with eyes that glow when you stroke its back? Could you make a burglar alarm? Try a bigger switch so that a foot makes it go on, and use thicker cardboard. If you put the switch under a model road layout, toy cars could turn it on – or make up a target game like this one!

Tip

If your switch stays on, try a smaller hole.

A simple pressure pad game

Some ideas to get you started

One way to use your pressure pad (see page 12) is to make a simple board game. A treasure hunt, for example, could lead to the magic spot by a series of clues, and a light or buzzer would come on when a player landed in the right square.

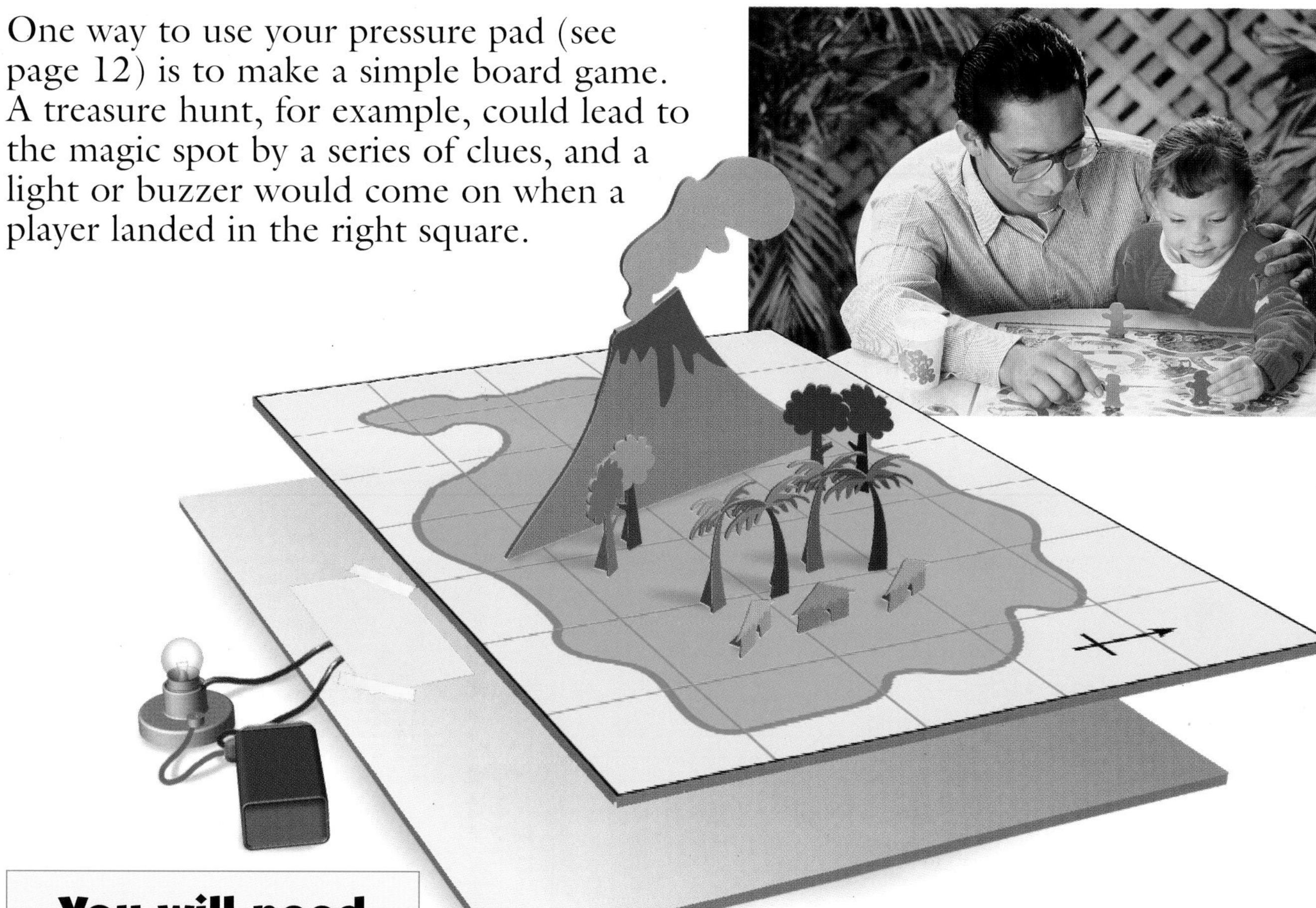

You will need

- a pressure pad with light or buzzer connected (see page 12)
- 2 sheets of thin cardboard
- masking tape or colored PVC tape
- coloring pencils and markers

You could also use:

- words and pictures done on a computer
- rub-on letters
- colored stickers

What to do

1. Make a pressure pad with a light or buzzer connected (see page 12).

2. Attach the pressure pad to one of the sheets of cardboard. Use masking tape or PVC tape so that you can move the pad to different places. How could this make the game more interesting?

3. Before adding your game design to the top layer of cardboard, find the sensitive spot and mark it with a pencil. You may need to add longer wires to your switch if making a large game.

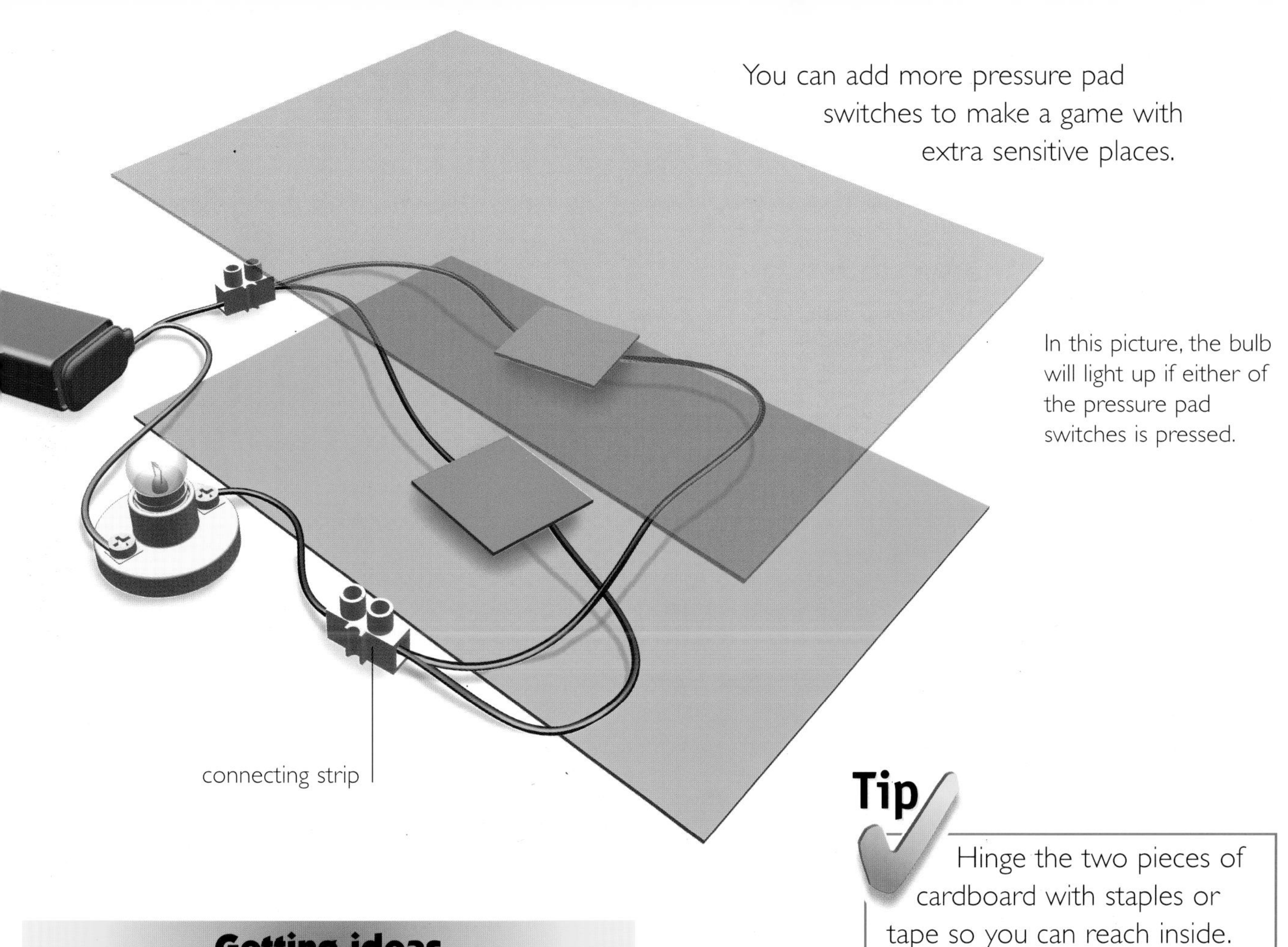

You can add more pressure pad switches to make a game with extra sensitive places.

In this picture, the bulb will light up if either of the pressure pad switches is pressed.

Tip

Hinge the two pieces of cardboard with staples or tape so you can reach inside.

Getting ideas

You could use a pressure pad switch in a game. Look at games that already exist. Talk to older people about games they used to play. Find out about games from different countries. All this might help you to design your own games – but don't just copy!

Think about who will play the game. How difficult should it be to play? Try to make up rules that are fair and easy to understand. If you can, use a computer to type out the rules.

Could you make counters or other extras? You might use painted pebbles, cardboard figures, or salt dough shapes. You could make packaging and advertising for your game.

More pressure pad games

Pressure pads with several sensitive places

Pressure pad panels on cash machines, microwave ovens, and other products often have lots of sensitive places.

Here is how to make your own pressure pad with several sensitive places. When any of these places is pressed, your circuit will be turned on. You can use your pad to make more complicated games.

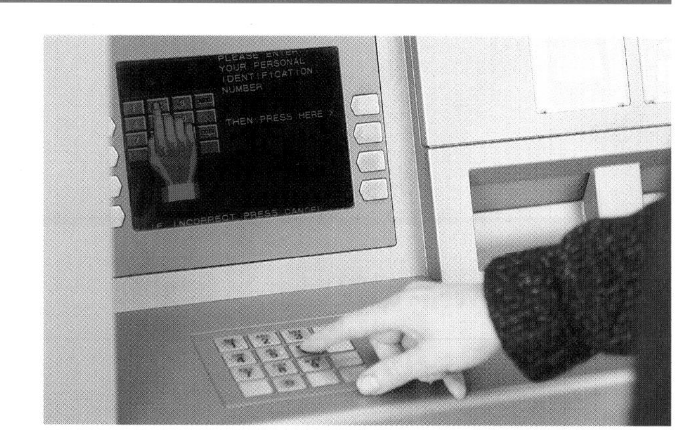

You will need

- 3 8.5" by 11" sheets of thin cardboard
- craft knife, safety ruler, and cutting board

Safety note: Ask an adult before using a craft knife.

- kitchen foil
- glue stick
- circuit to switch on
- stapler

What to do

1. Draw a grid on one sheet of cardboard, spacing the lines 3/4 inch apart. (Use faint pencil lines, as you may not want the lines to show later.) This sheet will be the top layer.
2. Cover the underside of the top layer with foil, except for a strip at one end.
3. Look at "Getting ideas" (page 17) and design your game or quiz. Decide which squares you want to be switches and mark them in pencil. You don't have to choose the ones in the picture – choose as many as you want.
4. Draw the same grid on the middle layer. Mark the same squares as you marked on the top layer and cut them out.

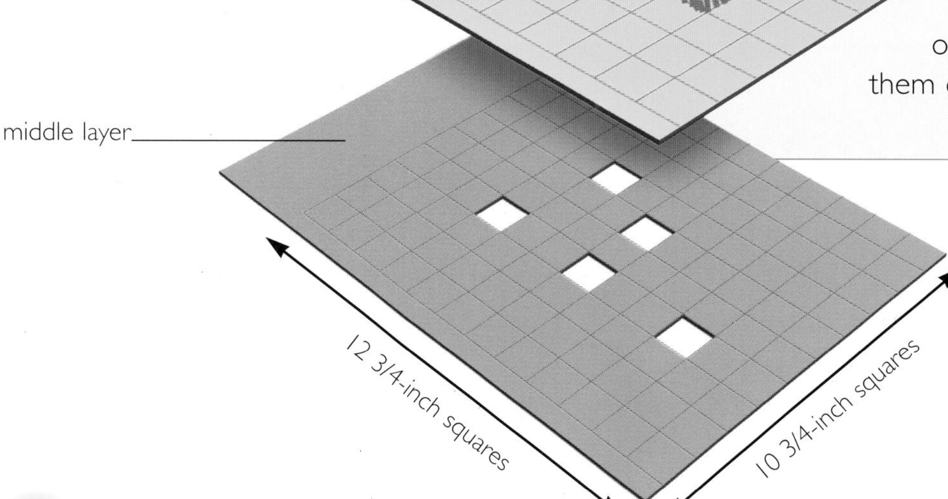

5. Cover the bottom layer with foil, except for a strip at one end.

6. Staple on the circuit. Staple the pieces of cardboard together at the end so they open like a book.

Important: Staple through the bare cardboard, or the staples will conduct electricity and your switch will stay on!

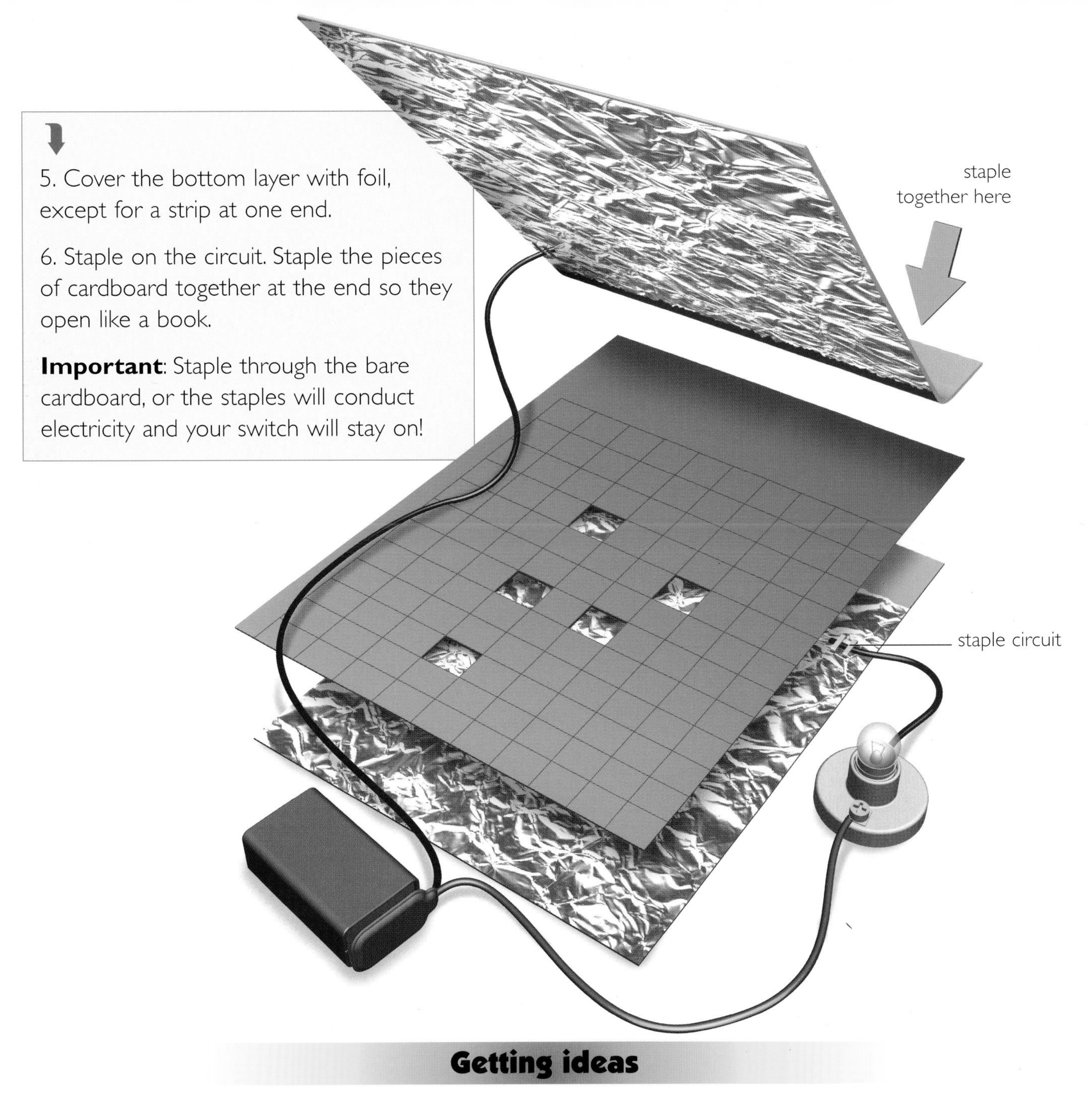

Getting ideas

Design a game using the pad you have made. You could look at games like Chutes and Ladders and design your own version. You could draw hazards (such as monsters) on the sensitive squares or leave them unmarked to surprise the players! The switch squares could show hidden treasure instead of hazards and dangers. Use your pad to design a quiz for younger children. Write or draw pictures showing the right answers on the sensitive squares. Show wrong answers on squares where there are no switches. By covering and uncovering some holes in the middle layer you could fool people who think they know where the switches are!

You could go on to design your own pads and think of your own uses for them.

A secret switch

A pressure pad with pairs of switches

Keypads are often used when only certain people are allowed to open a door. These people are told which numbers to press to open the door, and they keep the numbers secret. You can make a pressure pad where the circuit will only work by pressing the right two places at the same time.

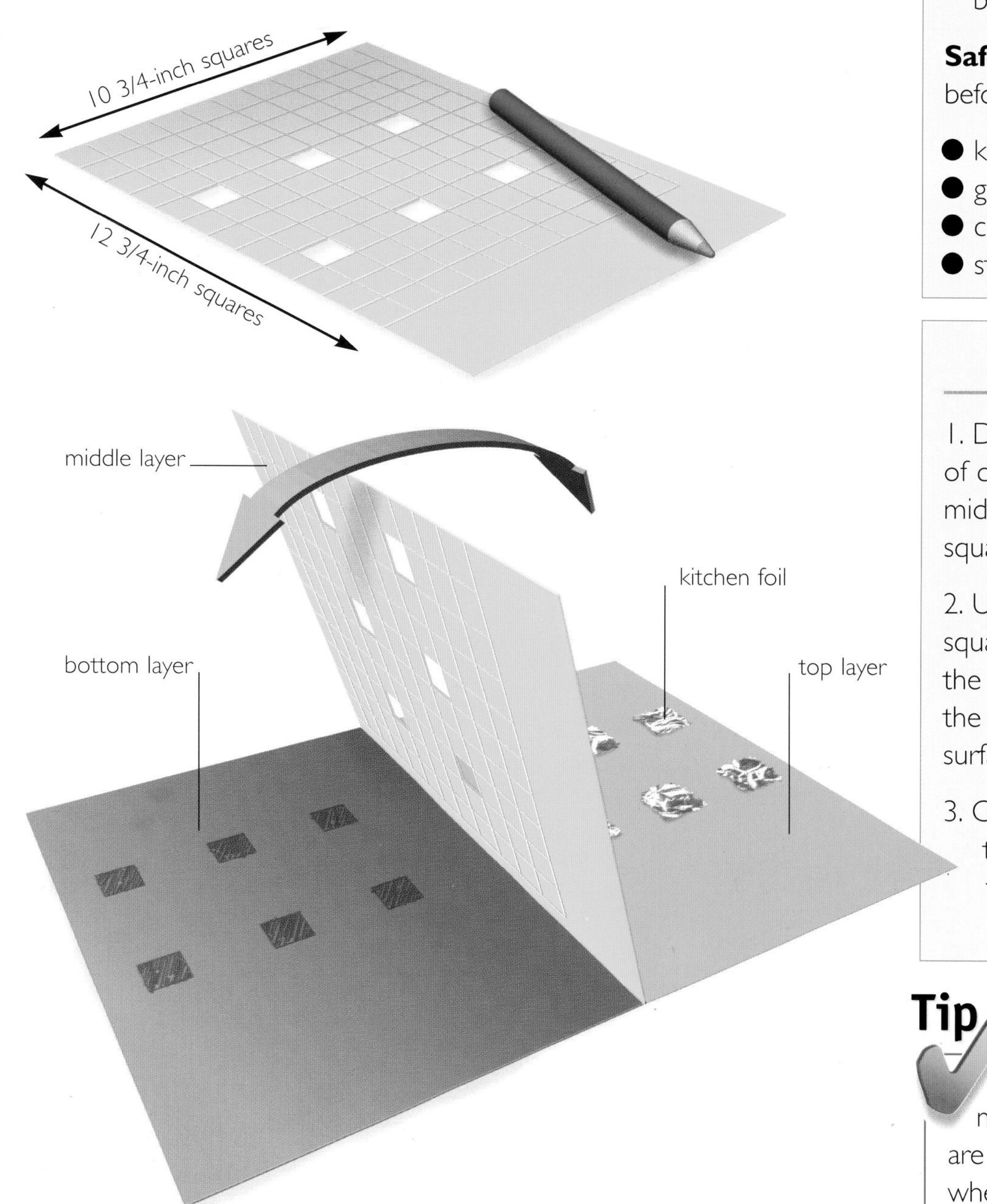

You will need

- 3 8.5" by 11" sheets of thin cardboard
- craft knife and cutting board

Safety note: Ask an adult before using a craft knife.

- kitchen foil
- glue stick
- circuit to switch on
- stapler

What to do

1. Draw a grid on one sheet of cardboard. This will be the middle layer. Cut out the squares shown.
2. Use the holes to stencil squares on the underside of the top layer. Also stencil the squares on the top surface of the bottom layer.
3. Cover the squares on the top layer with foil. Attach the foil with a glue stick.

Tip

Flip the "stencil" to make sure that the squares are one above the other when the pad is put together.

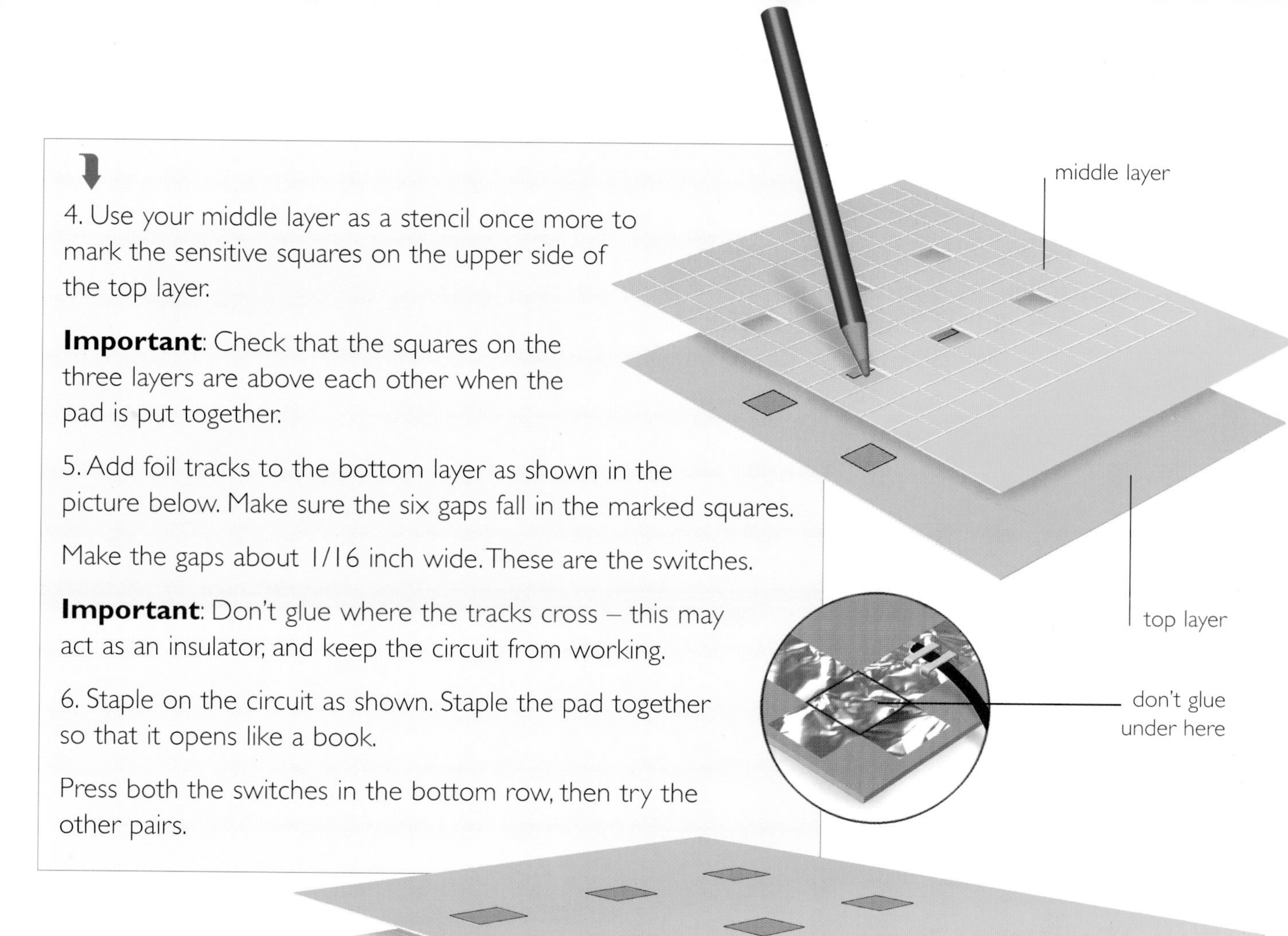

4. Use your middle layer as a stencil once more to mark the sensitive squares on the upper side of the top layer.

Important: Check that the squares on the three layers are above each other when the pad is put together.

5. Add foil tracks to the bottom layer as shown in the picture below. Make sure the six gaps fall in the marked squares.

Make the gaps about 1/16 inch wide. These are the switches.

Important: Don't glue where the tracks cross – this may act as an insulator, and keep the circuit from working.

6. Staple on the circuit as shown. Staple the pad together so that it opens like a book.

Press both the switches in the bottom row, then try the other pairs.

Getting ideas

You could design a quiz where two things have to be matched, such as animals with their homes, or sports players with their team or country. Put a pair of answers over each pair of switches. Put wrong answers over the rest of the sheet where there are no switches.

You could hide one of each pair of switches so that only you know how to turn the circuit on. Try to think of projects that use this idea.

Secret "combination lock" switch

Making your own security device

To make this switch work you have to turn the top layer to the right position and know where to press. Only those who know the secret can make the circuit work. All around you can see electronic ways of letting one person make something happen while stopping others. Your library card has its own special bar code. Credit and cash cards have their own magnetic strip and number. Special cameras can scan your eye to check that it's really you.

You can use this switch as a security device to decide who can turn on projects, or to make puzzles, games, and quizzes.

You will need

- 8.5" by 11" thin cardboard
- kitchen foil
- paper fastener
- circuit

What to do

1. Cut a 6-inch-wide strip of cardboard and fold it in half. Cut another piece to fit on top.

2. Cut a 1-inch-square "window" in the strip and use the hole to stencil a square on the other two layers.

3. Stick foil over the square on the top layer. Stick two strips across the bottom square. (They should be close together but not touching.)

4. Draw the largest circle that will fit on the top layer and cut it out.

5. Staple the circuit to the foil strips. Attach the top and middle layers with a paper fastener. Turn the circle to the right place and press!

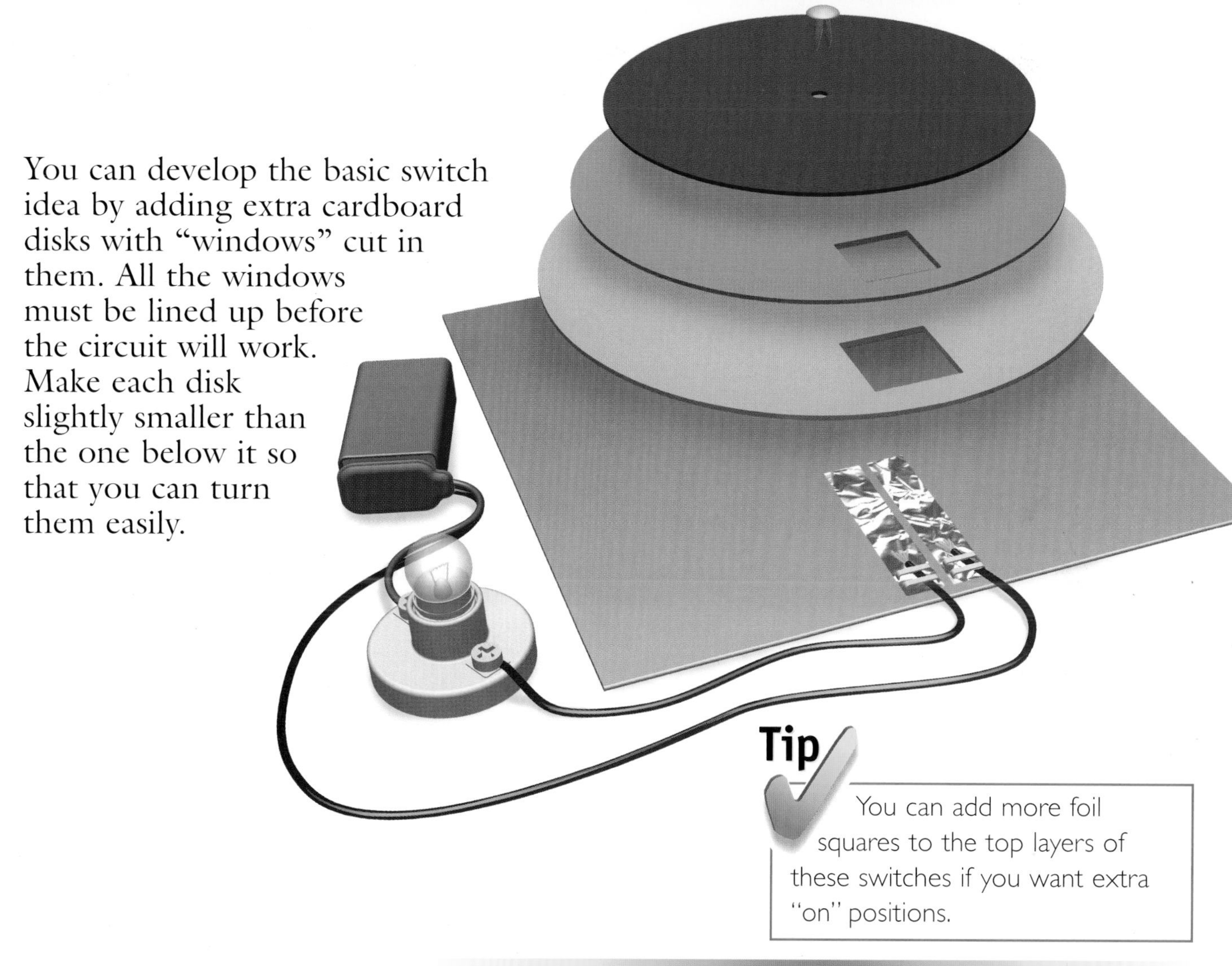

You can develop the basic switch idea by adding extra cardboard disks with "windows" cut in them. All the windows must be lined up before the circuit will work. Make each disk slightly smaller than the one below it so that you can turn them easily.

Tip

You can add more foil squares to the top layers of these switches if you want extra "on" positions.

Getting ideas

You could use the switch like a combination lock. Think of different things it could control. Could it control a motor that pulled back a catch? Could you make a "doorbell" that only friends would know how to use?

Think about using the switch in puzzles, games, and quizzes. You could put pictures around the disk. When the correct picture on the disk is lined up with its name, the switch would work. On the second version of the switch you could ask the player to line up all the pictures that go together before they press.

Try out your switch – it may help you think of other ideas.

The push-to-break switch

A switch that is always on – unless you press it!

A switch that is on all the time unless it is pressed is called a push-to-break switch because you push it to "break" (make a gap in) a circuit. You probably have one or more in your home. One use for this kind of switch is to control the light inside a refrigerator. When the door is shut it pushes on a switch, breaks the circuit, and keeps the light off, which saves energy. Here's how to make a push-to-break switch and other kinds of switches to use in design projects.

You will need

- piece of plastic bottle
- some kitchen foil
- glue stick
- stapler
- small piece of cardboard

A push-to-break switch
Press to switch OFF.

What to do

1. Cut two strips of plastic 1 inch wide and 3 inches long.

Stick foil to both sides using the glue stick.

2. Bend the strips into the shapes shown.

3. Stick the strips to the cardboard, making sure that they spring together until the bottom one is pressed.

4. Staple the circuit to the strips – it should stay on until you press the switch!

You can change the switch so that you have to press to turn the circuit on. Just pull one part above the other and you have a push-to-make switch. Can you see the difference between this switch and the first one?

A push-to-make switch
Press to switch ON.

In a knife switch, one part of the knife switch is turned to make contact with the other when you want a circuit to stay on.

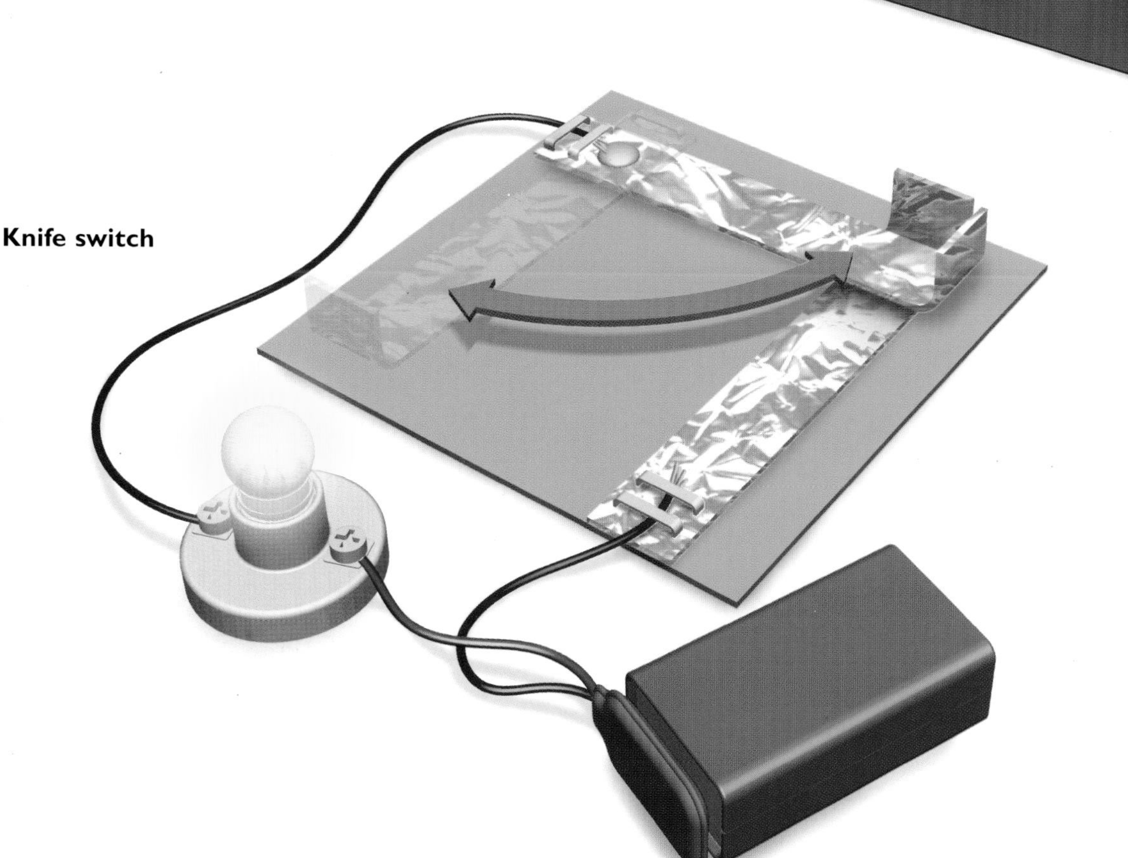

Knife switch

Getting ideas

Think of different ways to make the switches go on or off. You could leave something resting on the push-to-break switch. If anyone moves the object the circuit could warn you. Opening a lid or door could also make a circuit work.

The push-to-make switch could be used for signaling. You could use it to send messages in Morse code. You could use this switch as part of a pinball game or whenever you want a circuit to go on. How could you make the push-to-make switch stay on?

You could use the knife switch when one of your projects needs a circuit that will stay on, like the lights on a model vehicle. A cotton thread could pull the knife switch on.

The reed switch

A switch that works without being touched!

The reed switch works when a magnet is brought near it. Reed switches are used to sound an alarm when a burglar opens a door or window. They have also been used where the spark from an ordinary switch would cause an explosion. These switches can be hidden under cardboard, fabric, or other material – as long as the magnetism can reach them they will still work.

Only a magnet will make the switch work!

You will need

- reed switch
- magnet
- 2 pieces of connecting strip
- small screwdriver
- circuit

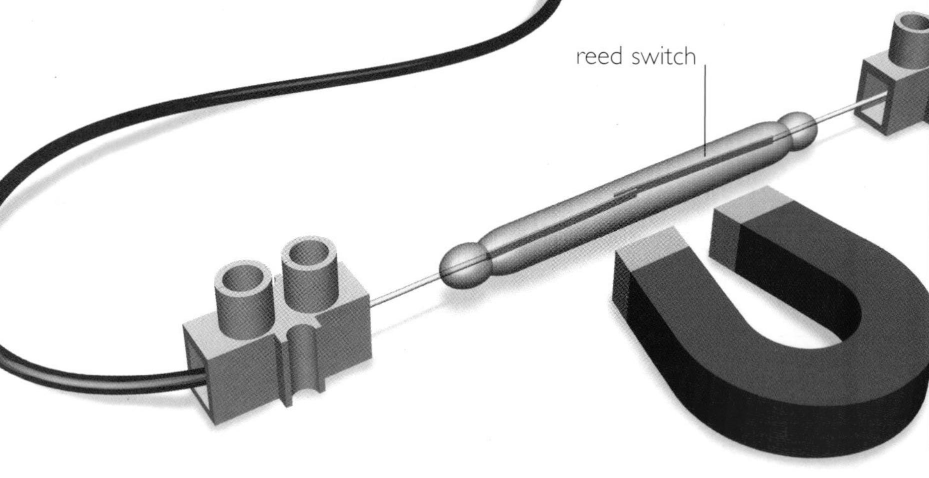

What to do

1. Use the screwdriver to attach a piece of connecting strip to each end of the reed switch.

Safety: Reed switches are made of glass, so handle them with care. Don't bend the wires or you may crack the switch.

2. Add a circuit to the other ends of the connector strips.

3. Bring a magnet close to the switch to test your circuit. You may need to turn the switch or magnet to make it work well.

A strong magnet will turn on the switch from some distance away – experiment!

To keep your reed switch in place on a project, you could use tape or plastic bag ties.

It works best to put the magnet on the moving part of a project instead of the circuit. What would happen on this project (right) if the wheel with the magnet turned faster?

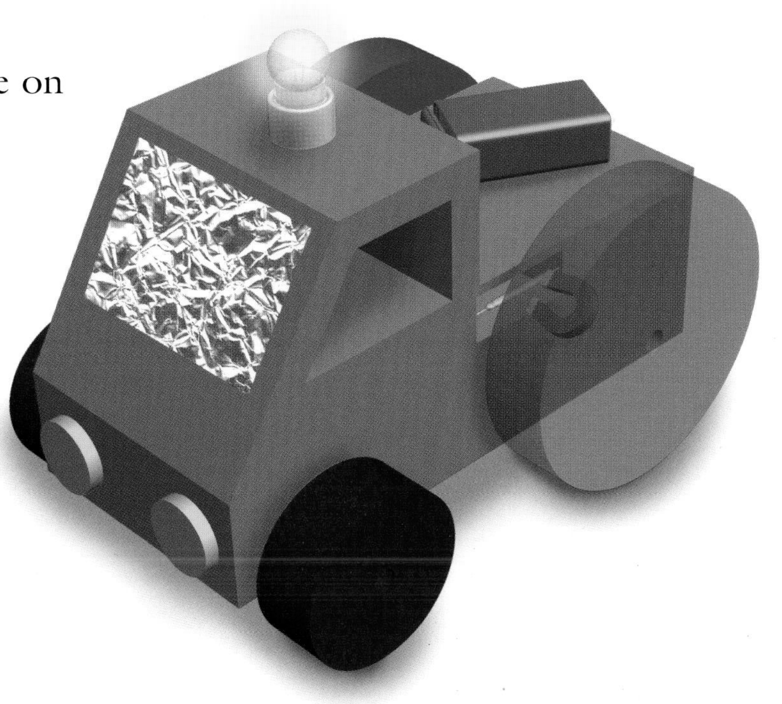

Tip

When using lights be careful not to leave the magnet near the switch by accident or you will waste your battery.

Getting ideas

There are many ways to move the magnet near the reed switch. It could be attached to a wheel, swung on a thread, slid in a tube, or stuck under a toy car. Think of other ways to bring the magnet close to the switch. Do any of these give you ideas for design projects?

Hiding the magnet and switch might give you ideas for design projects. The magnet could be hidden by attaching it to your finger with clear tape or putting it inside a magic wand. You could design a creature with eyes that lit up when only you stroked it. Remember, the reed switch is special because it doesn't have to be touched to make it work!

The tilt switch

A switch that works when something moves

When a ball rolls inside a tilt switch the circuit is turned on. A thief rocking a car could make a tilt switch set off an alarm. A pedometer (see photograph), which measures how far someone has walked, uses a tilt switch to count leg movements. Some tilt switches use a liquid metal called mercury. Others use a small metal ball. Here's how to make your own tilt switch and ideas to start you thinking about how to use it.

You will need

- 2 corks
- piece of plastic bottle to make the tube
- 2 paper clips (not colored ones)
- clear tape
- small glass marble
- kitchen foil

What to do

1. Make a tube 4 inches long by wrapping the plastic tightly around the corks. Hold the tube together with tape.
2. Wrap the marble tightly with foil to make a round shape that will roll easily.
3. Twist the bare ends of your circuit tightly around the paper clips.
4. Tape the paper clips to a cork. Make them stick well out from the cork. Push the cork and paper clips into one end of the tube.
5. Put the marble in the tube. Check that it moves easily and touches both paper clips – switching the circuit on! (You may need to bend the paper clips in a little.)
6. Once the switch works, tape both corks in place.

Tip

It is easier if you ask a friend to put the tape on while you hold the plastic around the cork.

The first switch works when it is tilted one way, which is fine for some projects. The switch shown here works when tilted either of two ways. Just add a second pair of paper-clip contacts.

You could try different lengths of tube. What difference do you think this would make?

The switch below works when tilted in any direction. What difference do you think different sizes of box would make?

Getting ideas

Use masking tape to attach the switch to things around you that move. Can you make the switch work when a door or one of your favorite things is moved? Can you hold the switch and climb over obstacles without setting it off?

Projects with moving parts could include the tilt switch to turn a circuit on automatically.

Think about using the tilt switches as starting points for design projects. The first switch could be built into a magic wand that lights up when pointed downwards at something. The box switch could be the start of a game design. Players could try to steer the ball around a track or maze without setting off an alarm.

A coin-operated switch

Make your own slot machine

Slot machines are used for lots of things, including selling candy, food, tickets, and soft drinks. You can buy things when the stores are closed by putting money in a slot. What else are slot machines used for? What could they be used for?

The switch shown here works when a metal coin conducts electricity and closes the gap in the circuit. You can see the gap between the strips of foil. Here is how to make your own switch so that a buzzer, light, or motor works when a coin is put in a slot. One use for a switch like this is to make collecting or saving money more fun. A well-designed money box might make people put more money in the slot just to make it work!

You will need

- a box (like a shoe box) or cardboard to make one
- a strip of corrugated cardboard
- kitchen foil
- a circuit (see pages 6-7)
- masking tape
- paper fasteners

kitchen foil

What to do

1. Choose (or make) a box that is big enough to hold your switch and collect the money.
2. Cut a strip of corrugated cardboard to the shape shown (above) and bend up the sides. The tubes inside the cardboard should run *along* the strip. Make the longest slide that will fit in your box.
3. Stick on two strips of foil using a glue stick. Smooth strips work best.
4. Staple the bare ends of a circuit tightly to the foil strips. You could add a capacitor (see page 9) to keep a light or buzzer on longer. Lay a coin on the switch to test it.

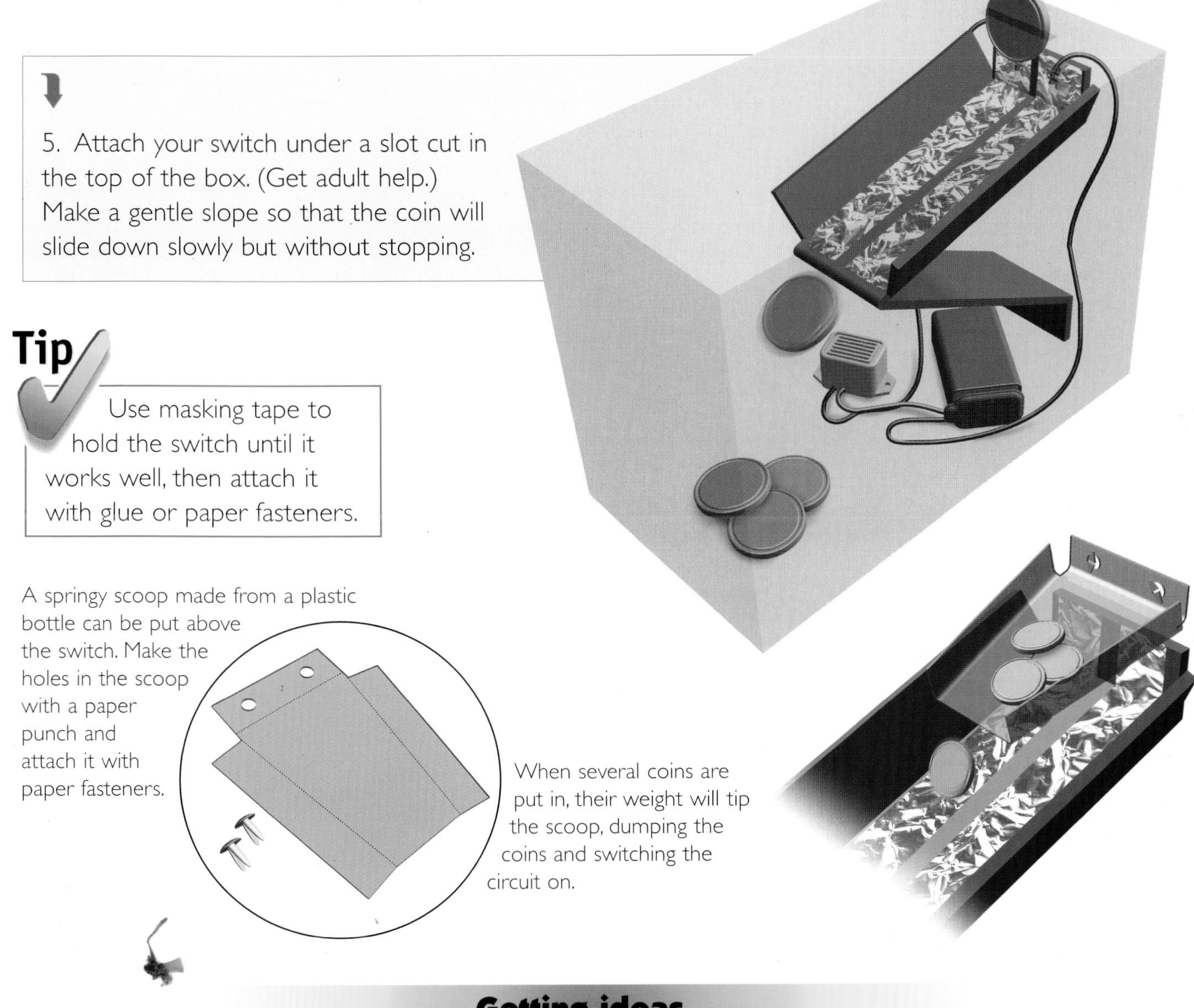

5. Attach your switch under a slot cut in the top of the box. (Get adult help.) Make a gentle slope so that the coin will slide down slowly but without stopping.

Tip

Use masking tape to hold the switch until it works well, then attach it with glue or paper fasteners.

A springy scoop made from a plastic bottle can be put above the switch. Make the holes in the scoop with a paper punch and attach it with paper fasteners.

When several coins are put in, their weight will tip the scoop, dumping the coins and switching the circuit on.

Getting ideas

You could design a money box where the switch turns on a light (see page 8). The light could be part of a picture on the box. A motor could make part of a picture spin. Clip-art pictures and word-processed writing from a computer could be used. You could design a box to collect money for a charity or other good cause.

The basic switch can be changed in different ways. A hole could be cut in the slide so that coins that are too small will drop through into a reject box without turning the circuit on. A long slide will keep your circuit on for a longer time than a short one. If the slide was held level, one coin would keep the circuit on all the time. Could this be useful for some projects? You could have a different switch to turn the circuit off. Your switch and coin collecting box could be part of a bigger project. Could it turn on a motor to pull back a catch so that a door could be opened?

Index